Sponsored Technology Workshops 2024

Held at the 2024 AIChE Spring Meeting and 20th Global Congress on Process Safety

New Orleans, Louisiana, USA
24-28 March 2024

ISBN: 979-8-3313-1720-1

Printed from e-media with permission by:

Curran Associates, Inc.
57 Morehouse Lane
Red Hook, NY 12571

Some format issues inherent in the e-media version may also appear in this print version.

AIChE
120 Wall Street, FL 23
New York, NY 10005-4020

Phone: (800) 242-4363
Fax: (203) 775-5177

www.aiche.org

Additional copies of this publication are available from:

Curran Associates, Inc.
57 Morehouse Lane
Red Hook, NY 12571 USA
Phone: 845-758-0400
Fax: 845-758-2633
Email: curran@proceedings.com
Web: www.proceedings.com

TABLE OF CONTENTS

72a PandID's and PSI/Maintenance Information Management: How Can They Work Together?

Stephen Nelson

An unfortunate trend in today's workflow is that once a P&ID has been released for construction, the integrity and data starts to get lost. As a manager you could pay upward of $2,000 for an updated P&ID and still end up with a piece of paper that does not achieve your end goals. In this workshop we can help you understand how to create a value-added P&ID that achieves ease of use and provides a PSI repository which in turn will create ownership for the PSI and PSM data at your facility and improve your auditing success. Most facilities information retrieval is sporadic and forces you down a winding rabbit hole of document accessibility and the path followed is usually very time-consuming and confusing. We can show you a P&ID with enhanced data storage capability to solve your information gaps and create successful information management retrieval. When information access and maintainability becomes simple, it causes your personnel to take ownership of the information management and its requirements. This workshop would benefit anyone that needs to access and maintain PSM PSI for their facility and wants to have a simple solution for information management.

28a The Digital SME for Knowledge Transfer in Operations

Clint Jeffus

Major customers from petrochemicals, specialty chem and refining have shared the challenges they face daily in their plants:

- Few SMEs (Subject Matter Experts), who are overwhelmed
- Turnover and onboarding
- Equipment specific incidents due to procedure execution and know-how
- Operational Discipline
- Training and regulatory requirements

This impacts process safety and other HSE/operations variables. We will dig deeper into a digital SME, available 24/7, where operators can review, practice procedures before they perform them. The platform can be used to qualify and requalify operators on equipment and units. We will look at the impact this has had in loading/unloading operations, technical gas plants, major petrochemical facilities, refineries and discrete manufacturing.

47a Dynamic Management of Safeguards - getting more and more granular

Alwin Van Aggelen

Showing the evolution of Dynamic Management of Safeguards

a. Started as (IOGP) barrier category + performance standard level mostly concentrated on maintenance data to inform barrier health

b. Moved into more specific monitoring with Dynamic Bowtie (monitoring critical scenarios and barriers on break-down element level). Adding more data than just CCMS data --> competency, sensors, alarm management etc.

c. Evolution of Dynamic Bowtie = System-based bowties that are closer reflecting the risk and status of processing systems

d. Next stage is even more detailed/granular with scenarios coming directly from PHA (HAZOP/LOPA) and visualizing impact of degradation on safeguards, while taking semi-quantitative approach (PFD/RRF) into account to show risk impact on dynamically updating risk matrix for current risk.

e. Process Safety Information (PSI) always available in digital form and (due to shared master data etc.) easily retrievable during PHA study prep

f. Operational information (MOCs, inspection results etc.) triggering PHA review under certain conditions

g. Future state = operational deviation trends feeding back to PHA study to make those fully dynamic and perform condition-based instead of time-based revalidations

71a Filling in Knowledge Gaps as a part of Process Safety Culture

Jeremy Pafford, Becky Pease

At the heart of process safety culture is the commitment of employers to their employees, who must be given the tools and opportunities to put process safety into action. One of the core challenges we have seen to making this commitment work is a lack of easy to use information resources for employees at all levels to understand and recognize safety issues. Minimizing the risk of knowledge gaps is one key area that organizations can address via the implementation of digital tools, resources libraries, and training regimens. This also helps organizations make the most of their workforce in managing process safety, shifting the culture from one where a small team handles all safety issues to one where every employee has the ability to recognize and report potential problems before they become larger issues. In this session we will describe how Elsevier has partnered with engineering firms to help nurture a process safety culture through the creation of easy to use tools that make it easier for employees at all levels to operate with process safety in mind.

27a Exploring the Invisible Threats in Pressure Relief and Flare Analysis: Bridging the Gaps

Nancy Faulk

Join us as we delve into real-world best practices, lessons learned, and case studies and learn from experts in the field. The workshop topics will extend through Pressure Relief and Flare Systems lifecycle from design, engineering, and data management. In addition, experts will present valuable lessons and common pitfalls that they encountered and explain how these oversights could lead to overpressure safety incidents if left undetected. Whether you're a seasoned professional or just starting your pressure relief journey, this workshop will provide valuable insights to enhance the effectiveness of your pressure relief and flare systems data management.

112a CCS and Green Hydrogen Process Modeling using Symmetry

Jordan Mangold

There is an increased focus on limiting global warming to below 2°C which requires a 140 times increase in installed carbon capture and storage (CCS) capacity. There are unique challenges when dealing with accurate modeling of the CCS process in the form of CO2 solidification, hydrates or even shifting between vapor and liquid boundaries. It is essential to evaluate these risks to operations and to recognize their impact on process downtime or performance. Increased interest in green hydrogen as a clean energy alternative requires the availability of process simulation tools that can model the full lifecycle from hydrogen generation to distribution. Hydrogen systems present challenges in terms of thermodynamics and safety systems, which may increase risk during design and operations. Please join this workshop to learn about how the Symmetry process simulation platform can help solve various design and operational challenges in CCUS and Green Hydrogen processes.

93a How to ensure your facility is safe in a cost-effective manner

Scott G. Davis

Facility owners and operators are often faced with difficult questions regarding their facilities' safety. Many of these decisions have serious economic repercussions in the lifecycle of an asset and require further engineering analysis. Retrofitting an existing facility, adding new equipment, or building a new module in a facility can represent millions of dollars in capital and operational expenditures, so it is imperative to conduct proper assessments to achieve safety objectives in a cost-effective manner. This technical workshop will highlight case studies on how to address safety challenges as they pertain to facility siting. We'll explore examples that demonstrate how to use quantitative, consequence-predictive tools for informed decision-making with safety as a priority. It will also address strengths and limitations with various study types as they increase in complexity, establishing a process to deliver comprehensive safety assessments with business economics in mind.

NOTES

NOTES

AUTHOR INDEX